I0465640

About the Author

Edilberto "Sandy" Santiago Jr. is a songwriter and blogger from Norzagaray, Bulacan, whose music celebrates honesty, kindness and community spirit. Raised in a humble agricultural family, he writes practical guides on farming techniques and survival skills tailored to the Philippine environment.

Passionate about arts and music, Edilberto uses his voice and his pen to inspire others—whether through uplifting verses or hands-on tips for life in the fields and the wild.

Permissions & Inquiries
Upper COC. Norzagaray
Bulacan 3002. Philippines
Email: edilberto@edilbertosantiago.com
Web: www.edilbertosantiago.com

TABLE OF CONTENTS

UNDERSTANDING CROP DIVERSIFICATION

Importance of Crop Diversification

Crop diversification is a strategic approach that can significantly enhance the productivity and sustainability of smallholder farming systems. For individuals working with less than one acre of land, diversifying crops is not just a method of increasing yield; it is a vital practice that mitigates risks associated with climate variability, pests, and market fluctuations. By growing a variety of crops, smallholders can create a more resilient farming system that can better withstand challenges. This method not only improves food security for the household but also stabilizes income throughout the year, as different crops have varying harvest times and market demands.

One of the primary benefits of crop diversification is the improvement of soil health. Different crops contribute differently to the soil, with some enhancing nutrient levels while others help prevent erosion and suppress weeds. For example, legumes are known for their ability to fix nitrogen in the soil, which can benefit subsequent crops when planted in rotation. This practice reduces the need for synthetic fertilizers, making farming more sustainable and cost-effective. Additionally, a diverse cropping system can attract beneficial insects and other wildlife, promoting natural pest control and enhancing biodiversity on the farm.

Diversification also opens up new market opportunities for smallholder farmers. By growing a range of crops, farmers can cater to various consumer preferences and reduce reliance on a single crop, which can be particularly risky if that crop fails or market prices fall. For instance, while staple crops like maize or rice may provide a reliable base,

4

incorporating high-value crops such as vegetables or herbs can significantly boost income. This strategic approach to crop selection allows smallholders to tap into niche markets, enhancing their economic viability and supporting local food systems.

Furthermore, crop diversification plays a crucial role in climate adaptation. As climate change continues to alter growing conditions, farmers face increased uncertainty regarding which crops will thrive. By cultivating a variety of crops, smallholders can better hedge against these uncertainties. Certain crops may perform well in drier conditions, while others may be more resilient to heavy rains. This flexibility not only helps in managing risks associated with climate extremes but also encourages innovative practices that can lead to improved agricultural outcomes.

Finally, the social and community benefits of crop diversification cannot be overlooked. When smallholders engage in diverse cropping practices, they often share knowledge, skills, and resources with their neighbors, fostering a sense of community and collaboration. This sharing can lead to collective problem-solving and enhanced resilience at the community level. Furthermore, diversified farming systems can promote cultural practices and traditional crops that are vital for maintaining local heritage and food sovereignty. In conclusion, crop diversification is not merely a technique for maximizing yield; it is a holistic approach that supports ecological balance, economic stability, and community cohesion for smallholder farmers.

Benefits for Smallholder Farmers

Smallholder farmers stand to gain numerous advantages from implementing crop diversification strategies on their limited plots of land. One of the most significant benefits is increased resilience against pests and diseases. By growing a variety of crops rather than a single staple, farmers can reduce the risk of total crop failure. This diversification minimizes the impact of specific pests or diseases that may target certain crops, allowing farmers to maintain a more stable income even when facing agricultural challenges.

Another advantage of crop diversification is the improvement of soil health. Different plants have varying nutrient requirements and root structures, which can enhance soil structure and fertility. For instance, legumes can fix nitrogen in the soil, benefiting subsequent crops. By rotating and intercropping different species, smallholder farmers can reduce the need for chemical fertilizers and improve the overall quality of their land over time, leading to sustainable farming practices that benefit both the environment and the farmer's livelihood.

Economic resilience is also a critical benefit of crop diversification. By cultivating a mix of high-value and staple crops, smallholder farmers can tap into various markets and reduce their dependence on a single crop for income. This strategy can lead to improved cash flow throughout the year, as different crops have staggered harvest times and market demands. Farmers can capitalize on seasonal market prices and ensure a more consistent income stream, which is vital for their financial stability.

Furthermore, crop diversification can enhance food security for smallholder farmers and their communities. By growing a variety of crops, farmers can ensure a steady supply of food for their families, reducing reliance on external food sources. This self-sufficiency is crucial, especially in times of economic uncertainty or food shortages. Additionally, diverse diets resulting from varied crop production can improve nutrition, leading to better health outcomes for farmers and their families.

Lastly, engaging in crop diversification can foster community connections and knowledge sharing among smallholder farmers. As farmers experiment with different crops and cultivation techniques, they often engage with one another to share insights and experiences. This collaborative environment can lead to the development of local networks that support sustainable practices and help farmers overcome common challenges. By working together, smallholder farmers can enhance their productivity and resilience while contributing to the overall agricultural development of their communities.

Common Myths about Crop Diversification

Crop diversification is often surrounded by misconceptions that can deter smallholder farmers from adopting this beneficial practice. One common myth is that crop diversification requires significant financial investment and resources, making it unsuitable for smallholder farmers with limited budgets. In reality, crop diversification can be achieved with minimal investment. Smallholders can start by intercropping or rotating a few different crops that require similar inputs or can thrive in the same soil conditions. This method not only reduces risk but also increases the resilience of the farming system without imposing heavy financial burdens.

Another prevalent myth is that crop diversification complicates farm management. Many smallholder farmers believe that managing multiple crops is too labor-intensive and time-consuming. However, with proper planning and organization, diversification can simplify farming practices. For instance, crop rotation can enhance soil fertility and reduce pest pressure, leading to less need for chemical interventions. By scheduling planting and harvesting times strategically, farmers can streamline their activities, making crop management more efficient rather than complicated.

Some individuals may think that crop diversification is only beneficial for larger farms, assuming that small plots cannot produce enough variety to make a difference. This notion overlooks the potential for smallholder farmers to maximize their yield and improve food security through diversified cropping. Small plots can be effectively utilized to grow a range of crops that complement each other. For instance, planting legumes alongside cereals can improve soil health while providing a varied diet for the farmer's household. This approach not only maximizes land use but also supports local markets by supplying diverse produce.

A common misconception is that diversifying crops leads to decreased overall yields. Farmers often fear that focusing on multiple crops will dilute their efforts and reduce the productivity of each individual crop. However, research has shown that diversified cropping systems often yield higher total production compared to monocultures. This increase is primarily due to improved soil health, better pest management, and the spread of risk over different crops. By embracing diversification, smallholders can create a more stable and productive farming environment that enhances their overall yield.

Lastly, some believe that crop diversification is a recent trend and lacks historical significance. In fact, various cultures around the world have practiced crop diversification for centuries, recognizing its benefits in maintaining soil health and ensuring food security. Traditional polycultures have been the backbone of many agricultural systems, providing a sustainable way to manage land. By learning from these historical practices and adapting them to modern contexts, smallholder farmers can harness the advantages of crop diversification, ensuring their long-term success and sustainability in agriculture.

ASSESSING YOUR LAND
Soil Quality and Types

Soil quality and types play a crucial role in determining the success of any agricultural endeavor, especially for smallholder farmers who often work with limited resources. Understanding the characteristics of soil can empower individuals to make informed decisions about crop selection, planting techniques, and soil management practices. Healthy soil is not only essential for plant growth but also influences water retention, nutrient availability, and microbial activity, all of which contribute to higher yields and sustainable farming practices.

Different types of soil possess unique properties that affect their suitability for various crops. The primary soil types include sandy, clay, loamy, and silty soils. Sandy soil is well-draining but often lacks nutrients and moisture retention capabilities. Clay soil, on the other hand, retains water and nutrients but can become compacted and hard, making it difficult for roots to penetrate. Loamy soil, a balanced mixture of sand, silt, and clay, is considered ideal for most crops due to its ability to retain moisture while providing good drainage and nutrient availability. Silty soil, rich in nutrients and finer particles, can also support a variety of crops but may require careful management to prevent compaction.

Soil quality is determined by its physical, chemical, and biological properties. Key indicators of soil quality include pH levels, organic matter content, nutrient composition, and soil structure. The pH level of soil affects nutrient availability; most crops thrive in a slightly acidic to neutral pH range of 6.0 to 7.0. Organic matter is vital for enhancing soil fertility and improving water retention. Regularly adding compost or other organic amendments can significantly boost soil quality over time. Additionally, understanding the nutrient composition—such as nitrogen, phosphorus, and potassium levels—allows farmers to tailor their fertilization strategies based on specific crop needs.

Implementing crop diversification strategies can enhance soil quality and overall farm productivity. By rotating different crops, farmers can reduce nutrient depletion and break pest and disease cycles. Leguminous plants, such as beans and peas, can fix nitrogen in the soil, benefiting subsequent crops. Intercropping, or planting different crops in close proximity, can maximize space and improve soil health by promoting biodiversity. Cover crops, such as clover or rye, can also be used during fallow periods to prevent erosion, suppress weeds, and enhance soil structure.

In conclusion, understanding soil quality and types is fundamental for smallholder farmers looking to optimize their small plots of land. By recognizing the unique properties of their soil and implementing effective management practices, individuals can enhance their soil health, increase crop yields, and promote sustainable farming. Embracing crop diversification strategies not only improves soil quality but also supports resilience against environmental challenges, ultimately leading to a more productive and rewarding agricultural experience.

Climate Considerations

Climate considerations play a crucial role in the success of smallholder farming, particularly when it comes to crop diversification. Understanding local climate patterns—including temperature ranges, rainfall distribution, and frost dates—enables farmers to select the most suitable crops for their specific conditions. For small plots of land, this knowledge is essential, as it helps maximize yield and minimizes the risks associated with climate variability. By tailoring crop choices to the climate, smallholder farmers can achieve better results and ensure a more sustainable farming practice.

Microclimates within small plots can significantly influence crop growth. Factors such as elevation, proximity to water sources, and wind exposure can create variations in temperature and moisture levels. Farmers should take the time to observe their plots throughout the year, noting areas that receive more sunlight or those that retain moisture longer. This information can guide decisions on crop placement, allowing for optimal growth conditions. For instance, shade-tolerant crops can be planted in areas with less sunlight, while drought-resistant varieties can be prioritized in well-drained, sunny spots.

Seasonal variations are another important climate consideration. Understanding the timing of planting and harvesting is critical for maximizing productivity. Smallholder farmers can benefit from using a staggered planting approach, which allows for continuous harvests throughout the growing season. This strategy not only increases the diversity of produce available but also helps mitigate risks associated with climate fluctuations. By planting a mix of crops with varying maturation times, farmers can ensure that they have a steady supply of food, regardless of unexpected weather events.

Pest and disease pressure can also be influenced by climate. Warmer temperatures and increased humidity can lead to a surge in pest populations or disease outbreaks, making it essential for smallholder farmers to monitor their crops closely. Crop diversification plays a significant role in managing these threats. By growing a variety of crops, farmers can reduce the likelihood of widespread pest infestations and diseases, as many pests and pathogens are crop-specific. Additionally, implementing practices such as companion planting can enhance natural pest control and improve overall crop health.

Finally, adapting to climate change is crucial for smallholder farmers. As weather patterns become increasingly unpredictable, the resilience of farming systems must be prioritized. Farmers can build resilience by diversifying their crops, incorporating agroecological practices, and utilizing drought-resistant or heat-tolerant varieties. Education and access to resources, such as climate data and sustainable farming techniques, are essential for empowering smallholder farmers to make informed decisions. By considering climate factors in their farming practices, smallholders can cultivate a productive and sustainable agricultural system that thrives despite environmental challenges.

Water Availability and Irrigation Options

Water availability is a critical factor for smallholder farmers seeking to optimize crop production on plots of land less than one acre. In regions with limited rainfall, understanding the local water resources, including surface water and groundwater, is essential. Rainfall patterns can be unpredictable, making it vital to assess the reliability of water sources throughout the growing season. Farmers should consider the topography of their land, as well as the proximity to rivers, ponds, or underground aquifers, which can serve as supplementary water sources. Knowledge of seasonal variations in water availability allows smallholders to plan their planting schedules and crop choices more effectively.

Irrigation options available to smallholder farmers vary widely in complexity and cost. Simple systems such as bucket irrigation or gravity-fed furrows can be effective for small plots. These methods require minimal investment and can be easily managed by individuals with limited experience. For those looking to increase efficiency, drip irrigation offers a targeted approach that delivers water directly to the plant roots, reducing water waste and encouraging healthy growth. Drip systems can be installed on small plots and are particularly beneficial in optimizing water use, especially in arid regions where conservation is essential.

Rainwater harvesting is another viable option for smallholder farmers, especially in areas that receive significant rainfall during specific seasons. Collecting rainwater from roofs or surface runoff can provide a sustainable water source for irrigation, reducing dependence on external water supplies. Small storage tanks or barrels can be constructed to hold harvested water, ensuring that farmers have access to this resource during drier periods. Implementing rainwater harvesting systems not only enhances water availability but also promotes environmental sustainability by maximizing the use of natural resources.

Crop diversification strategies also play a crucial role in managing water resources effectively. By growing a variety of crops that have different water needs, smallholders can optimize irrigation practices and reduce overall water consumption. For instance, intercropping drought-resistant plants with those that require more water can create a balanced ecosystem that supports soil health while conserving water. Additionally, crop rotation can help maintain soil moisture levels and nutrient availability, which further supports diverse plant growth and minimizes the risk of crop failure.

Finally, education and capacity building are vital for smallholder farmers to make informed decisions about water management and irrigation practices. Workshops, community meetings, and local agricultural extension services can provide essential training on effective water conservation techniques and innovative irrigation systems. By sharing knowledge and experiences, farmers can learn from one another and adopt best practices tailored to their specific environmental conditions. Empowering smallholders with the right tools and information will not only enhance their water management capabilities but also contribute to the overall success of crop diversification efforts on small plots of land.

CHOOSING THE RIGHT CROPS
Assessing Market Demand

Assessing market demand is a crucial step for smallholder farmers looking to maximize their yields and profits through crop diversification. Understanding what consumers want can guide decisions on which crops to plant, ensuring that efforts align with market trends and local preferences. Smallholder farmers often have the advantage of being able to respond quickly to changes in demand, making it essential to stay informed about current market conditions and consumer behaviors. By conducting thorough market research, farmers can identify high-demand crops that may not require extensive resources or large plots, thus optimizing their production for profitability.

One effective method of assessing market demand is to engage with local markets and community networks. Farmers can attend farmers' markets, food co-ops, and local agricultural fairs to observe which products generate the most interest and sales. Interacting with consumers and fellow farmers can provide valuable insights into emerging trends and popular varieties. Additionally, establishing relationships with local restaurants and grocery stores can reveal specific preferences for certain crops, allowing farmers to tailor their planting strategies accordingly. This grassroots approach not only builds community ties but also enhances farmers' understanding of their target markets.

Analyzing demographic data can also play a significant role in assessing market demand. Understanding the characteristics of the local population, including age, income levels, and dietary preferences, helps farmers make informed decisions about which crops may be most successful. For example, if a significant portion of the local population is health-conscious, crops that cater to this trend, such as leafy greens and superfoods, may present lucrative opportunities. Conversely, if the area has a strong cultural inclination towards certain foods, such as specific vegetables or herbs, farmers can capitalize on this knowledge by growing those varieties.

Seasonality is another critical factor to consider when assessing market demand. Certain crops may have peak seasons when they are in high demand, while others may be less sought after at different times of the year. Smallholder farmers can maximize their productivity by strategically planning crop rotations and planting schedules that align with these seasonal trends. Additionally, extending the growing season through techniques such as succession planting or using high tunnels can allow farmers to offer fresh produce when competitors may have limited supply, thereby capturing a greater share of the market.

Finally, utilizing online resources and platforms can enhance market demand assessments. Online tools can provide data on food trends, pricing, and consumer preferences across various regions. Social media platforms also serve as a means of gauging public interest in specific crops, as farmers can observe discussions and trends that may not yet be reflected in traditional market analysis. By combining these digital insights with on-the-ground research, smallholder farmers can make well-informed decisions that will lead to successful crop diversification and, ultimately, greater sustainability and profitability in their agricultural endeavors.

Companion Planting Principles

Companion planting is a strategic approach that involves growing different plants in proximity to enhance growth, repel pests, and improve overall yields. This technique is particularly beneficial for smallholders who are working with limited space and resources. By understanding the principles of companion planting, individuals can create a more resilient and productive garden ecosystem. The foundation of these principles lies in the natural relationships between plants, which can be harnessed to maximize the potential of small plots.

One key principle of companion planting is the concept of mutual benefit. Certain plants can enhance each other's growth when planted together. For example, the classic combination of tomatoes and basil not only improves the flavor of the tomatoes but also deters pests like aphids and whiteflies. Similarly, legumes, such as beans and peas, can fix nitrogen in the soil, benefiting neighboring plants like corn and squash. By choosing compatible plants, smallholder farmers can naturally increase soil fertility and promote healthier crops.

Another important principle is pest management through companion planting. Many plants have natural properties that repel harmful insects or attract beneficial ones. Marigolds, for instance, are known to deter nematodes and other pests, making them an excellent companion for a variety of vegetables. Additionally, planting flowers like nasturtiums can attract pollinators and predatory insects, which help control pest populations. Implementing these companion planting strategies can significantly reduce the need for chemical pesticides, promoting a healthier environment and reducing costs for small-scale farmers.

Companion planting also plays a crucial role in optimizing space and maximizing yields in small gardens. By intercropping or layering plants, gardeners can utilize vertical space and create a diverse planting scheme. For example, tall plants like sunflowers can provide shade for shorter crops such as lettuce or spinach, allowing for a more efficient use of sunlight. This method not only increases the variety of produce grown but also helps in creating a microclimate that can be beneficial for all plants involved.

Finally, the principles of companion planting encourage biodiversity, which is essential for the long-term sustainability of smallholder farms. A diverse planting scheme can break pest cycles, enhance soil health, and promote a balanced ecosystem. By integrating a variety of crops and companion plants, smallholders can create a more resilient system that can better withstand challenges such as climate variability and pest outbreaks. Embracing the principles of companion planting not only enhances productivity but also fosters an environment where small-scale farmers can thrive sustainably.

Seasonal Crop Selection

Seasonal crop selection is a critical aspect of maximizing yield and ensuring a sustainable harvest for smallholder farmers. Understanding the unique growing conditions of each season allows farmers to choose crops that thrive in specific temperatures, soil conditions, and moisture levels. By rotating crops according to seasonal suitability, farmers can enhance soil fertility, reduce pest populations, and ultimately increase the diversity of their produce. This practice not only contributes to a more resilient farming system but also offers a variety of produce that can meet local market demands throughout the year.

In the spring, farmers can focus on cool-season crops such as lettuce, spinach, and peas. These crops are well-suited to the cooler temperatures and can be planted early in the season, providing a quick turnaround for harvest. Choosing fast-growing varieties can help smallholder farmers generate income early in the year, allowing them to reinvest in their farming operations. Additionally, these crops can be interplanted with slower-growing crops, taking advantage of available space and resources, which further enhances crop diversity on small plots.

As the temperatures rise in summer, it becomes essential to transition to warm-season crops. This is the ideal time to plant tomatoes, peppers, cucumbers, and squash, which thrive in warmer soil and air temperatures. Farmers should pay attention to the specific growing requirements of each crop, such as sunlight and water needs, to ensure optimal growth. Implementing techniques like mulching and drip irrigation can help manage soil moisture and reduce the risk of drought, making these warm-season crops more productive.

In the fall, smallholder farmers can prepare for a second crop cycle by selecting late-season varieties that can withstand cooler temperatures. Crops such as kale, Brussels sprouts, and root vegetables like carrots and beets can be planted to take advantage of the falling temperatures, allowing for harvesting even after the first frost. Additionally, this season provides an excellent opportunity for planting cover crops, which not only improve soil health but also help prevent erosion and suppress weeds. By carefully planning crop selection in the fall, farmers can maximize their output and maintain soil vitality during the off-season.

Winter presents unique challenges, but it also offers an opportunity for smallholder farmers to explore greenhouse growing or cold frames for certain crops. With the right setup, crops like spinach, garlic, and certain herbs can be cultivated, providing fresh produce even in harsh conditions. Seasonal crop selection during winter can not only extend the growing season but can also increase the farmer's resilience against market fluctuations. By diversifying crops throughout the year, smallholder farmers can create a more stable income and ensure a continuous supply of fresh produce for their families and communities.

PLANNING YOUR GARDEN LAYOUT
Space Optimization Techniques

Space optimization techniques are essential for smallholder farmers aiming to maximize productivity on limited land. With less than one acre available, effective utilization of every square foot can significantly impact yield and sustainability. Implementing strategies such as vertical gardening, intercropping, and container gardening allows farmers to diversify their crops while making the most of their available space.

Vertical gardening is an innovative approach that utilizes vertical structures to support plant growth. By growing crops upwards rather than outwards, farmers can increase their planting density without expanding their footprint. This method is particularly advantageous for high-yielding crops like tomatoes, cucumbers, and beans. Utilizing trellises, wall systems, or even repurposed pallets, smallholders can create a layered garden that not only saves space but also improves air circulation and reduces pest issues, leading to healthier plants and higher yields.

Intercropping is another powerful technique that enhances space efficiency while promoting biodiversity. By planting complementary crops in close proximity, farmers can make use of the different growth habits and root structures of various plants. For instance, pairing nitrogen-fixing legumes with nutrient-demanding vegetables can optimize soil health and fertility. This method minimizes competition for resources and maximizes the output from a small area. Additionally, intercropping can help control weeds and pests naturally, further reducing the need for chemical interventions.

Container gardening offers a versatile solution for those with limited land, allowing crops to be grown in pots, buckets, or other containers. This technique is particularly useful for urban environments or areas with poor soil quality. Containers can be placed strategically to take advantage of sunlight and can be moved as necessary to optimize growing conditions. Furthermore, container gardening makes it easier for smallholders to experiment with different crops and varieties, facilitating crop diversification and enabling farmers to respond quickly to market demands and preferences.

Lastly, proper spacing and crop rotation are critical elements of space optimization. Understanding the specific spacing requirements for each crop can prevent overcrowding and ensure that plants have adequate access to nutrients, water, and sunlight. Additionally, practicing crop rotation helps maintain soil health and reduces the risk of disease and pest buildup. By carefully planning the arrangement and timing of crop planting, smallholder farmers can create a sustainable and productive growing environment, ultimately leading to a more bountiful harvest from their small plots.

Vertical Gardening Solutions

Vertical gardening solutions are an innovative approach for smallholder farmers looking to maximize their crop yields on limited land. This method involves growing plants upwards rather than outwards, making efficient use of space and allowing for a diverse array of crops to flourish in a confined area. By utilizing vertical structures such as trellises, wall planters, and tiered gardening systems, individuals can cultivate not only traditional vegetables but also herbs, fruits, and flowers, thereby diversifying their produce and enhancing their overall harvest.

One of the primary advantages of vertical gardening is the ability to improve air circulation and light exposure for plants. In a horizontal garden, overcrowding can lead to increased disease risk and competition for sunlight. Vertical gardening mitigates these issues by allowing plants to spread out in a three-dimensional space. Crops such as tomatoes, cucumbers, and pole beans are excellent candidates for vertical growth, as they naturally vine and can be trained to grow upwards, freeing up ground space for other crops that do not climb.

Incorporating vertical gardening systems can also lead to improved water management and soil health. With raised beds and vertical planters, water can be more easily directed to the root zones of plants, reducing waste and ensuring that moisture levels are optimal. Additionally, vertical systems can facilitate better drainage, which is crucial for preventing root rot and other water-related issues. By carefully selecting companion plants that thrive in vertical arrangements, smallholders can create a balanced ecosystem that supports the health of the soil and the plants growing within it.

Moreover, vertical gardening can enhance the aesthetic appeal of small plots, turning them into vibrant, productive spaces. A visually pleasing garden can attract pollinators and beneficial insects, which are essential for crop diversification and increased yields. Utilizing colorful flowers alongside crops can not only beautify the garden but also serve practical purposes, such as attracting bees for pollination or deterring pests. This integration of aesthetics and functionality can make gardening a more enjoyable experience and encourage community involvement.

Finally, adopting vertical gardening solutions can also provide economic benefits for smallholder farmers. By diversifying crops in a limited space, individuals can cater to local markets with a variety of fresh produce, increasing their potential income. Additionally, vertical gardening can reduce the need for expensive inputs such as fertilizers and pesticides, as healthier plants are often more resilient to pests and diseases. By embracing this method, smallholders can create sustainable systems that not only yield a diverse array of crops but also contribute to their economic stability and environmental sustainability.

Intercropping Strategies

Intercropping strategies involve growing two or more crops in proximity on the same land area, which can maximize productivity and enhance resource use efficiency. For smallholder farmers with less than one acre, intercropping offers a viable solution to improve soil health, increase yields, and diversify income sources. By strategically selecting compatible crops, farmers can create a more resilient agricultural system that minimizes pests and disease pressures while optimizing space and nutrients.

One of the primary benefits of intercropping is the ability to take advantage of different growth habits and resource requirements among crops. For example, pairing legumes with cereals can be particularly advantageous. Legumes, such as beans or peas, fix nitrogen in the soil, which benefits companion crops like maize or sorghum that require higher nitrogen levels. This symbiotic relationship not only boosts the overall health of the soil but also increases the yields of both crops, making efficient use of the limited space available on small plots.

Another effective intercropping strategy is the use of trap cropping, where a more attractive crop is planted alongside a primary crop to lure pests away. For instance, planting sunflowers or marigolds near vegetables can draw aphids and other pests away from the main crops. This natural pest management technique can significantly reduce reliance on chemical pesticides, promoting a healthier ecosystem and ensuring the sustainability of farming practices. Furthermore, by diversifying crops, farmers can mitigate the risk of total crop failure due to pest infestations or adverse weather conditions.

Crop rotation within an intercropping system can further enhance soil fertility and reduce disease incidence. By alternating crops, farmers can disrupt the life cycles of soil-borne pathogens and pests that thrive on specific crops. For example, after a season of planting potatoes, a farmer might follow up with a legume crop, which not only enriches the soil but also helps break the cycle of diseases like blight. This strategy is particularly beneficial for smallholder farmers, as it allows them to maintain soil health without the need for expensive fertilizers or chemical treatments.

Finally, successful intercropping requires careful planning and management, including consideration of planting times, growth rates, and harvest schedules. Farmers should select crops that complement each other in terms of height, rooting depth, and nutrient requirements to avoid competition for resources. Additionally, maintaining a diverse range of crops can provide a steady income throughout the growing season, as different crops can be harvested at various times. By embracing intercropping strategies, smallholder farmers can effectively transform their small plots into productive, sustainable agricultural systems that yield diverse and nutritious crops.

SUSTAINABLE PRACTICES FOR SMALL PLOTS

Organic Farming Fundamentals

Organic farming fundamentally revolves around the principles of sustainability, ecological balance, and biodiversity. For individuals working on small plots of land, typically less than one acre, understanding these principles is essential for maximizing productivity while maintaining environmental integrity. Organic farming avoids synthetic fertilizers and pesticides, relying instead on natural inputs and processes. This approach not only enhances soil health but also promotes a diverse ecosystem that can support a variety of crops and beneficial organisms.

Soil health is a cornerstone of organic farming. Healthy soil is rich in organic matter, nutrients, and microorganisms that contribute to plant growth. Smallholder farmers can improve soil health by incorporating practices such as crop rotation, cover cropping, and composting. Crop rotation helps prevent the buildup of pests and diseases, while cover crops can enhance soil structure and fertility. Composting provides a sustainable source of organic matter that enriches the soil, enhancing its ability to retain moisture and nutrients.

Biodiversity plays a critical role in organic farming, especially for smallholders seeking to diversify their crops. By growing a variety of plants, farmers can create a more resilient system that is less vulnerable to pests and diseases. A diverse planting strategy can also improve pollination and support beneficial insects, which are crucial for many crops. Companion planting, where specific plants are grown together to enhance growth or deter pests, is a practical approach that small plot farmers can implement to maximize their yields.

Effective water management is another essential aspect of organic farming. With limited land and resources, smallholder farmers must utilize efficient irrigation methods to ensure crops receive adequate moisture without wastage. Techniques such as drip irrigation or rainwater harvesting can be particularly beneficial. Additionally, mulching can help retain soil moisture and suppress weeds, further contributing to the sustainability of the farming system. These practices not only support plant health but also save time and labor, making farming more manageable.

Finally, education and community involvement are vital for successful organic farming. Smallholder farmers can benefit from local networks, workshops, and cooperative societies that share knowledge and resources. Collaborating with others fosters a sense of community while providing access to new ideas and techniques. Continued learning about organic farming practices, market trends, and crop diversification strategies can empower individuals to adapt and thrive in their agricultural pursuits, leading to greater food security and resilience in their livelihoods.

Pest Management without Chemicals

Pest management without chemicals is an essential strategy for smallholder farmers seeking to maintain the health of their crops and the environment. This approach not only minimizes chemical residues in food but also promotes biodiversity and sustainability within the agricultural ecosystem. By employing a variety of natural techniques, farmers can effectively control pest populations, ensuring that their crops thrive even in constrained spaces.

One effective method in non-chemical pest management is the use of companion planting. This technique involves growing different crops in proximity to one another, which can deter pests and enhance crop productivity. For instance, planting marigolds alongside vegetables can repel nematodes and other harmful insects. Similarly, intercropping legumes with cereal crops can improve soil health and reduce pest pressures. By strategically choosing compatible plants, farmers can create a more resilient cropping system that naturally suppresses pest populations.

Another valuable strategy is the implementation of physical barriers. Row covers, nets, and traps can protect crops from pests without the need for chemical interventions. These barriers can prevent insects from accessing plants while allowing sunlight and rain to nourish them. For instance, using floating row covers can serve a dual purpose: protecting seedlings from frost and deterring pests like aphids and flea beetles. Additionally, hand-picking larger pests such as caterpillars or slugs can be a practical solution for small plots, allowing for immediate control without affecting the surrounding ecosystem.

Encouraging beneficial insects is also a key component of chemical-free pest management. By creating an environment that attracts natural predators, such as ladybugs, lacewings, and parasitic wasps, farmers can enhance the biological control of pest populations. Planting flowering species that provide nectar and pollen can draw these beneficial insects to the garden. A diverse habitat not only supports pest management efforts but also contributes to the overall health of the farm, making it more resilient to pest outbreaks.

Finally, implementing regular monitoring and maintenance practices is crucial for successful pest management.

Farmers should observe their crops frequently for signs of pest activity or damage, allowing for early detection and intervention. Crop rotation can also play a significant role in breaking pest life cycles and reducing infestations. By diversifying crop rotation plans and incorporating a mix of planting strategies, smallholder farmers can create a sustainable and productive farming system that thrives without relying on chemical pesticides. This holistic approach fosters a healthier environment while ensuring a bountiful harvest.

Soil Fertility and Crop Rotation

Soil fertility is a crucial aspect of successful crop production, particularly for smallholders working with limited land. Healthy soil provides the necessary nutrients that plants need to thrive, ensuring bountiful harvests. Soil fertility can be enhanced through various practices, including the addition of organic matter, use of cover crops, and the implementation of crop rotation strategies. Understanding the importance of soil health allows smallholder farmers to create a sustainable agricultural system that can yield high-quality produce year after year.

Crop rotation is a key strategy that can significantly improve soil fertility. By alternating the types of crops grown in a specific area over time, farmers can disrupt pest and disease cycles, reduce soil nutrient depletion, and enhance soil structure. Different plants have varied nutrient requirements and contributions; for instance, legumes can fix nitrogen in the soil, enriching it for subsequent crops. This practice not only promotes a balanced nutrient profile but also minimizes the need for chemical fertilizers, making it an environmentally friendly approach suitable for small plots.

In addition to enhancing soil fertility, crop rotation can lead to improved biodiversity and ecological balance within the farming system. By diversifying the crops planted, smallholders can attract beneficial insects and microorganisms that support plant health. This increased biodiversity can help to mitigate risks associated with pests and diseases, which is particularly important for small-scale farmers who may not have access to extensive pest management resources. Implementing crop rotation contributes to a more resilient farming ecosystem that can adapt to changing conditions.

Planning an effective crop rotation schedule requires knowledge of the specific crops being grown and their interactions. Smallholders should consider factors such as growth cycles, nutrient needs, and the potential for crop diseases. A well-thought-out rotation plan might involve planting nitrogen-fixing legumes one season, followed by nutrient-hungry crops like corn or tomatoes the next. This strategic approach not only maintains soil fertility but can also optimize yields, ensuring that the small plot is used to its fullest potential.

Ultimately, integrating soil fertility management with crop rotation offers smallholder farmers a pathway to sustainable productivity. By prioritizing soil health and employing diverse cropping strategies, individuals can enhance their food security and economic viability. As smallholders learn to manage their land effectively, they can achieve greater yields and resilience, paving the way for successful small-scale farming that benefits both their families and their communities.

MAXIMIZING YIELD
Efficient Use of Resources

Efficient use of resources is critical for smallholder farmers operating on plots of land less than one acre. Understanding how to optimize available resources not only leads to increased productivity but also enhances sustainability. By adopting practices that promote the efficient use of water, soil, and nutrients, smallholders can maximize their crop yields while minimizing waste and environmental impact. This section will explore various strategies that enable farmers to make the most out of their limited space.

One of the primary resources that smallholders must manage is water. Efficient water management can significantly improve crop growth and resilience. Techniques such as drip irrigation allow for targeted watering, minimizing evaporation and runoff. Additionally, rainwater harvesting systems can be employed to collect and store rainwater, providing a supplemental water source during dry periods. By using these methods, farmers can ensure that their crops receive the appropriate amount of water without overusing this precious resource.

Soil health is another crucial element in the efficient use of resources. Implementing practices such as crop rotation and cover cropping can enhance soil fertility and structure. These practices help to prevent soil erosion, improve nutrient cycling, and reduce the need for chemical fertilizers. Smallholder farmers can also benefit from composting organic waste, which enriches the soil and supports healthier plant growth. By prioritizing soil health, farmers can cultivate diverse crops that thrive on their small plots.

Nutrient management is essential for maximizing crop yields while minimizing costs. Understanding the specific nutrient requirements of different crops can guide farmers in choosing appropriate fertilizers and amendments. Utilizing soil tests can provide valuable information about nutrient levels and deficiencies. By applying nutrients based on these tests, farmers can avoid over-fertilization, reduce input costs, and minimize environmental harm. This targeted approach ensures that crops receive the nutrients they need for optimal growth.

Finally, integrating crop diversification strategies can lead to more efficient use of all available resources. By growing a variety of crops, smallholders can spread risk, reduce pest and disease pressure, and improve soil health. Companion planting, where compatible plants are grown together, can enhance resource utilization by maximizing space and sharing nutrients. This holistic approach not only boosts productivity but also fosters a resilient farming system capable of adapting to climate variability and market demands. Through careful planning and resource management, smallholder farmers can achieve impressive results on their small plots.

Harvesting Techniques for Increased Productivity

Harvesting techniques play a crucial role in maximizing productivity for smallholder farmers. Understanding the timing and method of harvest can significantly influence the quality and quantity of produce. For crops grown on small plots, it is essential to plan harvesting according to the growth cycles and maturation of different crops. Regular monitoring of the crops allows farmers to determine the optimal harvest time, ensuring that fruits and vegetables are picked at their peak ripeness, which not only enhances flavor but also extends shelf life.

One effective harvesting technique is known as staggered harvesting, which involves picking crops in multiple rounds rather than all at once. This method is particularly beneficial for crops that have a prolonged harvest window, such as beans and squash. By harvesting selectively, farmers can ensure that they are gathering only the most mature produce while allowing the younger fruits to continue growing. This approach not only maximizes yield but also spreads labor demands over time, making it easier to manage workloads throughout the growing season.

Utilizing the right tools for harvesting can also lead to increased efficiency and productivity. Simple hand tools, such as sickles, harvest knives, or pruners, can significantly reduce the time and effort required to gather produce. For small-scale operations, investing in quality tools is more beneficial than larger, mechanized equipment, which may not be practical for small plots. Proper tool maintenance, including regular cleaning and sharpening, ensures that they function optimally, thereby reducing the risk of crop damage during the harvesting process.

Post-harvest handling is another crucial aspect of the harvesting process that can impact overall productivity. Once harvested, crops should be handled with care to minimize bruising and spoilage. Implementing proper storage techniques, such as using crates or baskets that allow for air circulation, can help maintain the quality of produce. Additionally, considering the timing of harvesting in relation to market demands can enhance profitability. Farmers should aim to harvest when prices are favorable, which may require adjusting harvesting schedules based on local market trends.

Lastly, integrating crop diversification strategies into harvesting practices can contribute to increased productivity. By planting a variety of crops with different harvest times and requirements, smallholder farmers can create a more resilient farming system. This approach not only spreads the risk of crop failure but also ensures a continuous supply of fresh produce throughout the growing season. Farmers can also experiment with intercropping, where compatible crops are grown together, allowing for simultaneous harvesting and potentially improved yields. By adopting these techniques, individuals working with limited land can maximize their harvests and achieve sustainable farming success.

Post-Harvest Handling and Storage

Post-harvest handling and storage are critical components of successful crop production, especially for smallholder farmers who work on plots of less than one acre. Once crops are harvested, the way they are managed can significantly affect their quality, shelf life, and marketability. Proper handling ensures that produce retains its nutritional value and aesthetic appeal, which are essential for both home consumption and sale. Understanding the best practices for post-harvest management can help smallholders maximize their harvests and minimize losses.

The first step in effective post-harvest handling is the careful harvesting of crops. Farmers should use appropriate tools to avoid bruising or damaging the produce. For instance, using sharp knives or shears can help in cutting vegetables cleanly, while gentle handling during collection prevents physical injuries that can lead to spoilage. Timing is also crucial; harvesting crops at the right maturity stage ensures optimal quality. For example, leafy greens should be harvested in the early morning

35

when temperatures are cooler, reducing stress on the plants and extending their freshness.

Once harvested, crops should be cleaned and sorted to remove any damaged or diseased items. This process not only improves the overall quality of the produce but also reduces the risk of spoilage during storage. For smallholder farmers, sorting can be done manually, but it is essential to maintain hygiene standards. Washing produce in clean water and using suitable drying techniques can help in preventing microbial growth. Properly sorted produce is more appealing to consumers and can fetch higher prices in local markets.

Storage conditions play a pivotal role in prolonging the shelf life of harvested crops. Smallholder farmers often face challenges related to temperature and humidity control, which can lead to significant post-harvest losses. Utilizing simple storage solutions such as ventilated baskets, shaded areas, or even underground storage can help maintain optimal conditions. For perishable items like fruits and vegetables, it is beneficial to create a cool, dark environment to slow down the ripening process. Additionally, crop rotation in storage can prevent cross-contamination and spoilage.

Finally, smallholder farmers should consider the potential for value addition during the post-harvest phase. This can involve processing surplus produce into products like jams, pickles, or dried fruits, which can be stored for longer periods and sold at a premium. Engaging in such practices not only helps in managing excess harvest but also diversifies income streams, contributing to the overall sustainability of small-scale farming. By implementing effective post-harvest handling and storage techniques, smallholder farmers can significantly enhance their productivity and profitability, ensuring they make the most out of their limited land.

MARKETING YOUR PRODUCE
Identifying Target Markets

Identifying target markets is a crucial step for smallholder farmers looking to maximize their productivity and profitability on limited land. Understanding your potential customers enables you to tailor your crop selection, marketing strategies, and production practices to meet specific needs and preferences. By focusing on the right market segments, you can enhance your chances of success and ensure that your efforts yield the best possible returns.

To begin, it is essential to conduct market research to identify local demand for various crops. This can involve visiting farmers' markets, grocery stores, and local restaurants to observe which produce items are popular. Engaging with potential buyers through surveys or informal conversations can provide valuable insights into consumer preferences. Additionally, utilizing online resources and social media platforms can help gauge trends and identify gaps in the market that your small plot could fill, such as organic produce or specialty crops.

Once you have gathered information about potential markets, segmenting them based on characteristics such as demographics, buying habits, and preferences will help refine your approach. For instance, consider whether you want to target families seeking fresh vegetables, restaurants looking for unique ingredients, or community-supported agriculture (CSA) subscribers interested in seasonal produce. Each segment may have different requirements and expectations, which will influence your crop choices and marketing strategies.

It is also important to consider the seasonality of your target market. Different crops have varying harvesting times, and understanding when your chosen market demands certain produce can optimize your planting and sales schedules. For example, if you identify a high demand for tomatoes in the summer, you must plan ahead to ensure a timely harvest. By aligning your production with market needs, you can minimize waste and maximize sales opportunities.

Finally, developing strong relationships with your target market can significantly enhance your success. Networking with local buyers, joining community groups, and participating in local events can create a loyal customer base and foster trust. Consistent communication with your market can also provide feedback, allowing you to adapt your strategies as preferences and trends evolve. By nurturing these connections, smallholder farmers can ensure that their diversified crop efforts produce sustainable profits and contribute to their community's food system.

Direct Sales vs. Cooperatives

Direct sales and cooperatives represent two distinct pathways for smallholder farmers to market their produce, each with its own set of advantages and challenges. Direct sales involve farmers selling their products directly to consumers, which can include farmers' markets, roadside stands, or community-supported agriculture (CSA) programs. This model allows farmers to establish a direct connection with their customers, setting their own prices and retaining a larger share of the profits. For smallholders growing diverse crops on less than one acre, direct sales can offer flexibility and the opportunity to build a loyal customer base that appreciates the quality and uniqueness of their produce.

On the other hand, cooperatives are collective organizations formed by farmers to pool resources, share knowledge, and market their products more effectively. By joining a cooperative, smallholder farmers benefit from economies of scale, reduced costs for inputs, and a stronger bargaining position when negotiating prices with buyers. Cooperatives can also provide access to shared facilities, such as processing plants or storage units, which can help smallholders manage their diverse crop outputs more efficiently and reduce post-harvest losses. For those engaged in crop diversification, cooperatives can serve as a vital support system, enabling farmers to navigate the complexities of producing and selling multiple types of produce.

The choice between direct sales and cooperatives often depends on individual circumstances, including the types of crops grown, market access, and personal preferences. For farmers focused on niche markets or specialty crops, direct sales may provide the best opportunity to maximize profit margins. This approach allows them to highlight the quality and uniqueness of their products, catering to consumers who are increasingly interested in locally sourced and sustainably grown food. Conversely, farmers who may not have the time or resources to market their products independently may find cooperatives to be an advantageous option, as they can rely on the collective efforts of their peers to reach a wider audience.

Moreover, the decision can also be influenced by the local agricultural landscape and community dynamics. In areas where farmers' markets are thriving and consumer demand for local produce is high, direct sales can be a lucrative avenue. However, in regions where smallholders face competition from larger agricultural enterprises or where access to markets is limited, cooperatives can provide essential support by pooling resources and strengthening market presence. Understanding the local context and market dynamics is crucial for smallholders when deciding between these two models.

Ultimately, both direct sales and cooperatives can play significant roles in the success of smallholder farmers practicing crop diversification. Each model offers unique opportunities and challenges that can align differently with individual goals and circumstances. Farmers must weigh the benefits of direct consumer engagement against the support and collective power offered by cooperatives. By thoughtfully considering their options, smallholder farmers can enhance their profitability and sustainability, ensuring that their small plots yield not only abundant harvests but also economic resilience.

Building Brand and Community Connections

Building a strong brand and fostering community connections are essential components for smallholder farmers looking to thrive in a competitive agricultural landscape. A well-defined brand can communicate the unique value of your produce, while strong community ties can enhance your market reach and create a loyal customer base. By focusing on these areas, smallholder farmers can not only improve their economic viability but also strengthen their local agricultural networks.

Establishing a brand begins with identifying what sets your farm apart from others. This could be the specific crops you grow, your farming methods, or your commitment to sustainability. For instance, if you specialize in heirloom vegetables or organic practices, make this a central theme in your branding efforts. Consistency in your messaging across all platforms—be it social media, farmers' markets, or local grocery stores—helps consumers recognize and remember your brand. Utilize storytelling to share the journey of your farm, highlighting the care that goes into each crop and the positive impact of local farming on the community.

Community connections can be cultivated through active participation in local events and organizations. Attend farmers' markets, agricultural fairs, and community gatherings to network with potential customers and fellow farmers. Offering workshops or farm tours can also engage the community, allowing people to learn about your farming practices while fostering a sense of belonging. Collaborative efforts with local businesses, such as restaurants or food co-ops, can create partnerships that benefit both parties and promote local agriculture, further embedding your brand within the community.

Social media platforms provide an invaluable opportunity for smallholder farmers to build their brand and connect with the community. Share regular updates about your farming activities, seasonal crops, and any events you are participating in. Engaging content, such as recipes using your produce or behind-the-scenes glimpses of your farm, can spark interest and encourage followers to visit your farm or purchase your products. By creating an online community, you not only reach a wider audience but also foster a sense of connection with those who appreciate local food systems.

Finally, consider the importance of feedback in strengthening your brand and community ties. Encourage customers to share their experiences and suggestions, whether through surveys, social media, or direct conversations. This feedback can provide valuable insights into customer preferences and help you adapt your offerings to better meet their needs. By demonstrating that you value their input and are willing to make changes, you enhance customer loyalty and build a stronger, more resilient brand that is deeply rooted in the community.

CASE STUDIES AND SUCCESS STORIES

Examples of Successful Smallholder Diversification

Successful smallholder diversification often emerges from innovative practices and adaptive strategies that enhance productivity and resilience. One notable example is the case of a smallholder farmer in Kenya who transformed a half-acre plot into a thriving micro-farm. By integrating vegetables, fruits, and herbs, this farmer created a polyculture system that maximizes land use and reduces pest pressures. Crops such as kale and spinach were interspersed with tomatoes and peppers, allowing for staggered harvests and continuous cash flow. This diversification not only improved food security for the farmer's family but also provided surplus for local markets, demonstrating the economic benefits of varied crop production.

Another inspiring example comes from a smallholder in India who incorporated agroforestry into his farming practices. By planting fruit trees such as mango and guava alongside traditional crops like rice and legumes, this farmer was able to create a more sustainable agricultural system. The trees provided shade, improved soil fertility through leaf litter, and served as a windbreak, which enhanced the overall productivity of the land. The diversified system not only protected the crops from extreme weather but also generated additional income from fruit harvests, illustrating how integrating trees with crops can lead to greater resilience and profitability.

In Southeast Asia, a smallholder farmer adopted a strategy of aquaponics, combining fish farming with vegetable cultivation in a small plot. By utilizing the nutrient-rich

44

water from fish tanks to irrigate crops such as lettuce and herbs, this farmer achieved remarkable results. The closed-loop system minimized water use and eliminated the need for chemical fertilizers, while also providing a source of fish protein for the family. This innovative approach showcased how smallholders could leverage technology and ecological principles to create productive, sustainable farming systems that thrive on limited land.

A smallholder in Latin America exemplified the advantages of seasonal crop rotation by growing diverse crops throughout the year. This farmer strategically planted beans, maize, and squash in succession, optimizing soil health and reducing the risk of crop failure due to diseases or pests. By rotating these crops, the farmer maintained soil fertility and reduced the need for synthetic inputs, thus promoting sustainability. This practice not only ensured a steady supply of food but also allowed for greater income stability, emphasizing the importance of understanding seasonal dynamics in smallholder farming.

Lastly, a community-based project in West Africa highlighted the collective benefits of crop diversification among smallholders. Farmers engaged in a cooperative model, sharing resources and knowledge while diversifying their crops. By growing a mix of cassava, yams, and various vegetables, they were able to mitigate risks associated with market fluctuations and climate variability. The cooperative approach enhanced their bargaining power, improved access to markets, and fostered a sense of community resilience. This example illustrates how collaboration among smallholders can amplify the benefits of diversification, leading to greater overall success and sustainability in agriculture.

Lessons Learned from Failures

Failures in agriculture are often viewed as setbacks, but they can also serve as valuable learning experiences for smallholder farmers. Understanding what went wrong can inform future decisions and strategies. Small plot farmers, in particular, can benefit from analyzing their failures as they navigate the complexities of crop diversification. By examining past mistakes, farmers can develop resilience and adaptability, ensuring that their subsequent efforts lead to greater success.

One common failure among smallholders is the selection of inappropriate crops for their specific environment. This can stem from a lack of research or reliance on trends rather than understanding local conditions. For example, planting a crop that requires extensive moisture in a dry region can lead to crop failure. Farmers can learn from such experiences by conducting thorough research on soil types, climate conditions, and water availability before choosing which crops to diversify. This ensures that they select species that are not only profitable but also sustainable in their specific context.

Another frequent pitfall is inadequate planning and resource management. Smallholders may underestimate the time, labor, and inputs required for diverse crop production. This often results in overextending their capabilities, leading to poor crop management and ultimately, failure. By taking a lesson from this, farmers can implement better planning practices, such as creating detailed schedules and budgets. This careful planning allows them to allocate their resources more efficiently and leads to more successful crop rotations and intercropping strategies.

Pest and disease management is another critical area where failures can provide essential lessons. Smallholders may neglect to implement integrated pest management strategies, resulting in significant crop losses. Learning from such failures involves recognizing the importance of proactive measures, such as planting pest-resistant varieties, rotating crops, and using organic pest control methods. By adopting these strategies, farmers can minimize the impact of pests and diseases, leading to healthier crops and better yields.

Lastly, failures in marketing and distribution can severely impact the profitability of smallholder farms. Many farmers may produce surplus crops but struggle to find reliable markets or fair prices. The lesson here is the importance of building relationships with local buyers, exploring value-added options, and understanding market demands. Engaging in community-supported agriculture or farmers' markets can provide better access to customers and improve profitability. By learning from past marketing missteps, farmers can create a more sustainable income from their diverse crops, ensuring their long-term success.

Innovative Approaches from Around the World

Innovative approaches to crop diversification have emerged globally, offering smallholder farmers new avenues for maximizing their yields and enhancing food security. In regions facing climatic challenges, such as sub-Saharan Africa, farmers are increasingly adopting intercropping methods. This practice involves growing two or more crops in proximity, which can significantly improve soil health, increase biodiversity, and reduce pest infestations. For instance, the combination of maize and beans is widely practiced, as the beans fix nitrogen in the soil, benefiting the maize. Such strategies not only optimize land use but also provide a safety net against crop failure, ensuring a more resilient agricultural system.

In Southeast Asia, the concept of agroforestry is gaining traction as a sustainable approach to crop diversification. Farmers integrate trees with their annual crops, creating a multi-layered system that enhances productivity and biodiversity. The shade provided by trees can improve microclimates for crops, reduce soil erosion, and increase water retention. Moreover, agroforestry systems can produce additional income sources, such as fruits and nuts, which can be harvested alongside traditional crops. This holistic method not only boosts yields but also contributes to the long-term health of the ecosystem, making it a viable option for smallholder farmers looking to diversify their produce.

Latin America showcases another innovative strategy through the practice of cover cropping. Farmers are planting cover crops during the off-season to improve soil fertility, control weeds, and prevent soil erosion. Crops such as vetch and clover are used to enhance soil organic matter and suppress pests. This practice has proven beneficial for smallholders, as it requires minimal investment and simultaneously prepares the land for subsequent cash crops. By improving soil health, cover cropping leads to better yields, which is essential for smallholders who often operate on limited resources and need to maximize their output.

In the Mediterranean region, smallholder farmers are embracing the concept of vertical farming within their limited plots. By utilizing vertical space, these farmers can grow a variety of crops, from herbs to vegetables, while minimizing land use. This method is particularly advantageous in urban settings, where land is scarce, allowing individuals to cultivate fresh produce in small areas. Innovations such as hydroponics and aquaponics are also being integrated into these systems, providing efficient water use and reducing the need for chemical fertilizers. As urban populations grow, these innovative practices offer a sustainable solution to food production in confined spaces.

Finally, the rise of community-supported agriculture (CSA) models is an important trend in crop diversification. This approach fosters direct relationships between farmers and consumers, allowing smallholders to grow a diverse range of crops tailored to local demand. By engaging with their communities, farmers can ensure a stable market for their produce, reducing the risks associated with crop failures. Additionally, CSAs encourage diversity in farming practices as they promote the cultivation of seasonal and heirloom varieties, contributing to a richer local food system. This model empowers smallholder farmers to innovate and adapt, leading to increased resilience and success in their agricultural endeavors.

OVERCOMING CHALLENGES
Dealing with Climate Change Impacts

Dealing with climate change impacts is crucial for smallholder farmers aiming to achieve sustainable yields from limited land. Climate change has introduced unpredictable weather patterns, increased pest prevalence, and altered growing seasons, all of which can significantly affect crop productivity. As such, smallholder farmers must embrace adaptive strategies to mitigate these impacts and ensure a reliable food supply. This can involve selecting crop varieties that are resilient to changing climatic conditions, such as drought-resistant or flood-tolerant species. By diversifying crops, farmers can reduce the risk of total crop failure, as different plants may respond uniquely to environmental stresses.

Implementing crop rotation is an effective strategy for managing soil health and preventing pest and disease cycles exacerbated by climate change. By alternating different types of crops, farmers can enhance soil fertility, minimize erosion, and disrupt the lifecycle of pests that thrive on specific plants. For instance, following a heavy feeder like corn with a legume can replenish nitrogen levels in the soil while providing a secondary source of income. This approach not only diversifies production but also builds resilience against the changing climate by ensuring that the soil remains productive over time.

Integrating agroforestry practices can also play a significant role in addressing climate change impacts. By planting trees alongside crops, farmers can create a microclimate that moderates temperature extremes and reduces soil erosion. The shade provided by trees can protect certain crops from excessive heat, while the root systems can help retain soil moisture during dry spells. Moreover, trees can provide additional resources such as

50

fruits, nuts, and timber, contributing to both biodiversity and economic resilience. This multifaceted approach helps smallholders adapt to climate variability while enhancing their overall farm productivity.

Water management is another critical aspect of dealing with climate change impacts. Smallholder farmers should consider adopting efficient irrigation techniques, such as drip irrigation or rainwater harvesting, to optimize water use. These methods not only conserve water but also ensure that crops receive the necessary moisture during dry periods. Additionally, planting cover crops can improve soil structure and increase water infiltration, helping to retain moisture in the ground. By managing water resources effectively, farmers can safeguard their crops against drought and ensure consistent yields throughout the growing season.

Finally, education and community collaboration play vital roles in addressing climate change challenges. Smallholder farmers can benefit from participating in local agricultural networks and sharing knowledge about crop diversification strategies. Workshops, extension services, and peer exchanges can empower farmers with the skills necessary to adapt to changing conditions. By working together, communities can develop collective solutions, such as seed banks for resilient varieties or shared irrigation systems, enhancing their ability to withstand climate-related adversities. Through collaboration and continuous learning, smallholder farmers can cultivate resilience and secure their livelihoods in the face of climate change.

Financial Management for Smallholders
Financial management is a crucial aspect for smallholders aiming to achieve success in crop diversification. Smallholders often operate with limited financial resources,

making effective management of their finances essential for sustainability and growth. Understanding key financial concepts and practices can help smallholder farmers make informed decisions, optimize their investments, and increase their overall productivity. From budgeting to record-keeping, these skills enable farmers to navigate the complexities of agricultural finance.

Budgeting is the first step in effective financial management. Smallholders should create a detailed budget that outlines expected income and expenses associated with diverse crops. This budget should account for costs related to seeds, fertilizers, labor, irrigation, and other inputs. By having a clear financial plan, farmers can prioritize their spending, allocate resources efficiently, and anticipate any cash flow issues. A well-structured budget allows smallholders to set realistic financial goals and track their progress toward achieving them.

Record-keeping is another vital element in financial management. Accurate records of income and expenses provide smallholders with insights into their financial performance. This practice helps in evaluating which crops are most profitable and which might require adjustments or elimination. Simple methods such as using notebooks or digital tools can facilitate effective record-keeping. Regularly reviewing these records allows farmers to make data-driven decisions, assess the impact of crop diversification strategies, and identify trends over time.

Understanding financing options is also important for smallholders. Many farmers may need additional funds to invest in new crops or expand their operations. Exploring various financing avenues, such as microloans, grants, or cooperative funding, can provide the necessary capital. Smallholders should research local financial institutions and agricultural programs that offer support tailored to their needs. Being informed about these options can open doors to resources that enhance their ability to diversify successfully.

Finally, smallholders must embrace financial literacy as an ongoing process. This involves continuously seeking knowledge about agricultural markets, investment strategies, and economic trends that affect their operations. Participating in workshops, engaging with agricultural extension services, and networking with other farmers can enhance financial acumen. By staying informed and adaptable, smallholder farmers can improve their financial management skills, leading to increased productivity and the potential for greater harvests from their small plots of land.

Accessing Resources and Support

Accessing resources and support is a crucial aspect for smallholder farmers looking to diversify their crops and maximize their yields on limited land. Various institutions, both governmental and non-governmental, offer a wealth of information, training, and funding opportunities tailored to small-scale agriculture. Understanding how to navigate these resources can significantly enhance a farmer's ability to implement effective crop diversification strategies. It is essential for individuals to identify local agricultural extension services, which often provide valuable guidance on crop selection, pest management, and soil health.

Community support networks play a vital role in the success of smallholder farmers. These networks can consist of local farmer groups, cooperatives, or associations that facilitate knowledge sharing and collective purchasing of seeds and supplies. By joining these groups, individuals gain access to experiences and insights from fellow farmers who have successfully diversified their crops. Such collaborations can lead to improved market access, as collective selling often increases bargaining power and reduces costs associated with marketing and distribution.

Financial support is another critical resource for smallholders. Many organizations offer grants, microloans, and technical assistance specifically aimed at improving agricultural practices. Small farmers should explore options such as government subsidies, international aid programs, and local credit unions that cater to agricultural needs. Understanding the application processes and eligibility criteria for these financial resources can empower farmers to invest in tools and techniques that will enhance crop diversity and increase productivity.

In addition to community and financial resources, educational opportunities play an important role in supporting smallholder farmers. Workshops, webinars, and online courses are increasingly available, providing practical training on innovative farming practices. Farmers can learn about sustainable techniques, organic farming, and integrated pest management, all of which contribute to successful crop diversification. Engaging with local agricultural universities and research institutions can also yield insights into emerging trends and practices that are specifically suited to small plots of land.

Finally, leveraging technology can significantly enhance access to resources and support for smallholder farmers. Mobile applications and online platforms provide farmers with real-time information on weather patterns, market prices, and best practices in crop management. By utilizing these technologies, farmers can make informed decisions that optimize their crop diversification efforts. Embracing digital tools not only facilitates access to information but also connects farmers with a broader network of experts and resources that can further aid in their agricultural journey.

THE FUTURE OF SMALLHOLDER FARMING

Trends in Crop Diversification

As smallholder farmers seek to optimize their limited land resources, crop diversification has emerged as a key strategy for enhancing productivity and sustainability. This approach involves cultivating a variety of crops within a small area, which can lead to improved soil health, reduced pest pressures, and increased resilience to climate change. By diversifying crop types, farmers can create a more balanced ecosystem that supports beneficial insects and microorganisms, ultimately leading to healthier plants and higher yields.

One of the most significant trends in crop diversification is the integration of traditional and indigenous crops into farming systems. These crops are often well-adapted to local conditions and may require fewer inputs than more widely cultivated varieties. Growing indigenous crops not only helps preserve genetic diversity but also strengthens food sovereignty by allowing communities to rely less on externally sourced seeds and inputs. Additionally, these crops can provide unique nutritional benefits, appealing to consumers who are increasingly interested in diverse, healthy diets.

Another trend gaining traction among smallholder farmers is the practice of intercropping, where two or more crops are grown simultaneously in the same field. This method can maximize space and resources, as different crops may use nutrients and water differently, thereby reducing competition. For example, planting legumes alongside cereals can enhance nitrogen fixation in the soil, which benefits both crops. Intercropping not only boosts yield per unit area but also improves overall resilience against pests

56

and diseases, as the presence of multiple crops can disrupt pest life cycles.

Agroforestry is also becoming a popular trend in crop diversification, as it incorporates trees into agricultural landscapes. This practice can provide shade, shelter, and additional income through fruit or timber production, while simultaneously improving soil structure and fertility. The integration of trees into small plots can enhance biodiversity and create microclimates that benefit various crops. Furthermore, agroforestry systems can sequester carbon, contributing to climate change mitigation efforts while enhancing farmers' livelihoods.

Finally, market-oriented diversification is gaining importance as smallholder farmers look to increase their income. By growing a mix of high-value crops such as specialty vegetables, herbs, and flowers, farmers can tap into niche markets and consumer demand. This approach not only helps diversify income sources but also reduces financial risk by spreading dependency across multiple crops. With access to markets and suitable marketing strategies, smallholder farmers can transform their small plots into profitable enterprises, ultimately achieving greater food security and economic stability.

Technology and Innovation in Small Farming

Technology and innovation play a crucial role in enhancing the productivity and sustainability of small farming operations. For individuals working with less than one acre of land, the adoption of modern agricultural technologies can lead to significant improvements in crop yield and resource management. Tools such as drip irrigation systems, solar-powered water pumps, and soil moisture sensors enable smallholder farmers to optimize water usage and ensure that crops receive the appropriate amount of moisture. These technologies can minimize waste and lower costs, ultimately resulting in a more efficient farming practice.

Moreover, precision agriculture, a method that uses various technologies to monitor and manage field variability, can be incredibly beneficial for small plots. By utilizing GPS technology and data analytics, farmers can make informed decisions about planting, fertilizing, and harvesting. This targeted approach allows for the efficient allocation of resources based on the specific needs of different crops, making it easier to implement crop diversification strategies. For instance, understanding soil variability within a small plot can help farmers select the right combination of crops that thrive in specific areas, thereby maximizing productivity.

Innovative tools, such as mobile applications and online platforms, have also transformed how smallholder farmers access information and market their produce. These technologies provide farmers with real-time data about weather patterns, pest management, and market prices, enabling them to make timely and informed decisions. Online marketplaces allow farmers to connect directly with consumers, eliminating intermediaries and ensuring they receive fair prices for their products. By leveraging these digital platforms, small farmers can expand their reach and diversify their income sources, which is essential for financial stability.

The integration of sustainable practices through technology is another vital aspect of modern small farming. Techniques such as vertical farming and hydroponics are gaining popularity among smallholders, allowing them to maximize production in limited spaces. These innovative methods not only conserve water and land but also reduce the need for harmful pesticides and fertilizers. By adopting environmentally friendly technologies, small farmers can enhance their productivity while contributing to the health of the ecosystem, fostering a more sustainable approach to agriculture.

Finally, collaboration and knowledge-sharing among smallholder farmers can significantly amplify the benefits of technology and innovation. Community networks, workshops, and cooperative farming models encourage farmers to exchange ideas and experiences related to new technologies and practices. By learning from one another, smallholders can collectively enhance their farming techniques, improve crop diversification, and increase their resilience to challenges such as climate change and market fluctuations. Embracing a culture of innovation and collaboration will empower small farmers to thrive in an increasingly competitive agricultural landscape.

Building Resilience in Smallholder Systems

Building resilience in smallholder systems is crucial for enhancing the sustainability and productivity of farming practices on limited land. Smallholder farmers often face numerous challenges, including climate variability, pest infestations, and market fluctuations. By adopting strategies that promote resilience, farmers can ensure more stable yields and income, ultimately leading to greater food security and community well-being.

One effective approach to building resilience is through crop diversification. Growing a variety of crops rather than a single staple can reduce risks associated with crop failure. For instance, if one crop is affected by disease or adverse weather, others may still thrive, providing a buffer against total loss. Additionally, diverse cropping systems can improve soil health, enhance nutrient cycling, and promote beneficial insect populations, all of which contribute to a more robust farming ecosystem.

Intercropping and companion planting are practical methods that smallholder farmers can implement to increase biodiversity on their plots. By strategically planting different crops in proximity, farmers can optimize land use and enhance yields. For example, planting legumes alongside cereals can improve soil fertility through nitrogen fixation while maximizing space and resources. This not only supports the farmer's food needs but also contributes to ecological balance and resilience.

Furthermore, integrating agroforestry practices can significantly bolster the resilience of smallholder systems. Incorporating trees into farming practices provides additional benefits such as shade, windbreaks, and improved soil moisture retention. Trees can also yield fruits, nuts, or timber, offering supplementary income streams for farmers. This multi-layered approach to farming creates a more stable environment, mitigating risks associated with climate change and market shocks.

Finally, fostering community ties and knowledge-sharing can enhance resilience in smallholder systems. By collaborating with local farmers, individuals can learn from each other's experiences, share resources, and access collective markets. Community-supported agriculture (CSA) models can facilitate direct sales to consumers, ensuring a stable income for farmers. Emphasizing resilience through crop diversification not only empowers individual farmers but also strengthens the entire agricultural community, paving the way for a sustainable future.